Bibliographic information published by the German National Library:

The German National Library lists this publication in the National Bibliography; detailed bibliographic data are available on the Internet at http://dnb.dnb.de .

Imprint:

Copyright © 2018 GRIN Verlag
Print and binding: Books on Demand GmbH, Norderstedt Germany
ISBN: 9783668783997

This book at GRIN:

https://www.grin.com/document/437812

Amanuel Kussia

Transforming Konso towards Green Economy through Integrated Land Management

GRIN Verlag

Transforming Konso towards Green Economy through Integrated Land Management

By: Amanuel Kussia

August, 2018

Table of Contents

List of Tables

Table 1: Trend of Population Growth, Population Density, and Land Holding

Table 2: Occurrence of Drought in Konso

List of Figures

Figure 1: Location of Konso in Ethiopia

Figure 2: The DPSIR Framework for Land Resource Assessment

Figure 3: Annual Rainfall Anomalies…

Figure 4: Scene of Some of the Degraded Areas

Figure 5: Delibena River during Dry and Rainy Season

Figure 5: Segen River during Dry and Rainy Season…

Abbreviations

DPSIR	Drivers, Pressures, State, Impacts and Responses
FAO	Food and Agriculture Organization of the United Nations
IEA	Integrated Environment Assessment
KDA	Konso Development Association
KSWARDO	Konso Special Woreda Agriculture and Rural Development Office
LADA	Land Degradation Assessment
UNCCD	United Nations Convention to Combat Desertification
UNEP	United Nations Environment Program
UNESCO	United Nations Educational, Scientific and Cultural Organization

Abstract

This article examines the drivers, pressures, and impacts of land degradation on the ecosystem services and livelihood of the Konso people. To deal with the problem of land degradation, the People of Konso have been practicing well organized and innovative adaptation strategies in the form of indigenous soil and water conservation. The people are well known for their indigenous knowledge and skills of land management. Particularly, the antique and beautiful terraces, traditional agro forestry practices, and efficient irrigation methods are the prominent features of the Konso agricultural system. The people have been a model for the global community and their cultural landscape was registered by the UNESCO as one of the world's heritage sites. The indigenous soil and water conservation practices have enabled the community to manage and live in a callous natural environment. However, these practices are now under the threat due to multiple factors which requires the attention of all stakeholders, principally of the Konso people (that is primarily responsible to maintain its identity of soil and water conservation).

1. Introduction

Land is one of the few 'precious finite resources' Brabant (2010:4) upon which all living organisms depend for survival and growth. Particularly, in developing countries where the majority of the population lives in rural areas and entirely depends on land, people can neither exist nor prosper without this resource. Unfortunately, due to 'human activities which are exacerbated by natural processes, and often magnified by the impacts of climate change and biodiversity loss UNCCD (2013:4), the quality of land has been declining in an alarming rate. As a result, land degradation is becoming a horrifying event all over the world in general and developing countries in particular. For example, Bai, et al. (2008) reported that at global level, the percentage of total land area already degraded or being degraded increased from 15% in 1991 to 25% in 2011. Likewise, FAO (2011) reported that up to 25% of all land is highly degraded and 36% is slightly or moderately degraded. More than ever, the effect of land degradation on both human well-being and ecosystem services is highly devastating. For instance, Nachtergaele, et al. (2010) indicated that land degradation directly affects 1.5 billion people around the world and has already reduced the productivity of the world's terrestrial surface by about 25% from 1981 to 2003.

When it comes to Africa, the continent is extremely affected by land degradation and constitutes for 65% of world's arable land degradation (Thiombiano and Tourino-Soto, 2007). The World Bank cited in (Ibid) indicated that no less than 485 million Africans are adversely affected by land degradation. Land degradation is also one of the serious environmental problems facing Ethiopia and hampers the country's effort of poverty reduction and sustainable development. Given that the country is highly dominated by agriculture, the problem not only declines agricultural productivity which adversely affects livelihoods of its people and local economies but also leads to reduction in biodiversity and stream sedimentation that affect water quantity and quality, storage, and marine resources (Jolejole-Foreman, Baylis, and Lipper, 2012). Some studies conducted so far (Girma, 2001; Temesgen, Amare, and Hagos, 2014) indentified rapid population growth, severe soil loss, deforestation, low vegetative cover and unbalanced crop and livestock production as the major causes of land degradation in Ethiopia.

The people of Konso are well known for their indigenous knowledge and skills of land management. Particularly, the antique and beautiful terraces, traditional agro forestry, and efficient irrigation methods are the prominent features of Konso agricultural system. These practices have enabled them to conserve soil and water and live in extremely difficult environment. Even if the Konso people have such astonishing knowledge and skills of protecting their natural resources (land, soil, water, and forest), land degradation has been one of the serious problems that threaten the livelihoods of the people and practice of soil and water conservation. Many people including myself have a key question to ask: 'If these people have rich and innovative soil and water conservation practice and become model for the whole world, why has land degradation been their critical problem'? This essay tries to answer this and other related questions. In other words, the main purpose of this paper is to identify the drivers, pressures, and impacts of land degradation on the ecosystem services and livelihood of the Konso people.

2. A Brief Background about Konso Woreda

Konso is one of the five woredas[1] of the Segen Area Peoples' Zone of Southern Nations, Nationalities and Peoples Regional State (SNNPRS) located at a distance of about 595 and 360 km as of Addis Ababa and Hawassa, respectively. Konso woreda has 44 peasant associations and two towns. Total area of the woreda is estimated to be 2354 km^2 with a total population of 234, 987 where males and females constitute 48.3% and 51.7% of the total inhabitants, respectively (CSA, 2007). Topography of the woreda is characterized by hilly and mountainous intersected by valleys, gullies, ragged and plains. The landscape of the woreda is classified as 10% mountainous, 60% hilly undulating, and 30% plain and flat (Konso Special Woreda Agriculture and Rural Development Office (KSWARDO), 2008). Agro-climatically, the woreda is classified as 70% Kola (low altitude below 1500) and 30% Woinadega (mid altitude above 1500). The rainfall is of bimodal and erratic in nature of which 83% falls from February to May referred as Belg rain and 17% from August to November (Mehere, minor rain season) (Konso Development Association (KDA), 2003). In Konso, household economy is based on subsistence mixed farming: crop and livestock production. Sorghum is major crop staple food followed by maize and pulse crops. Other crops grown in smaller quantity include teff, millet, wheat, barley, haricot bean, and pigeon pea. Livestock species such as cattle, goat, sheep and chicken are kept by Konso farmers as a secondary livelihood. There are two cropping seasons: Belg (February-May) accounting for 65-75% of the annual crop production and Meher (*Hagayata*).

---

[1] Woreda is the third-tier administrative divisions of Ethiopian government

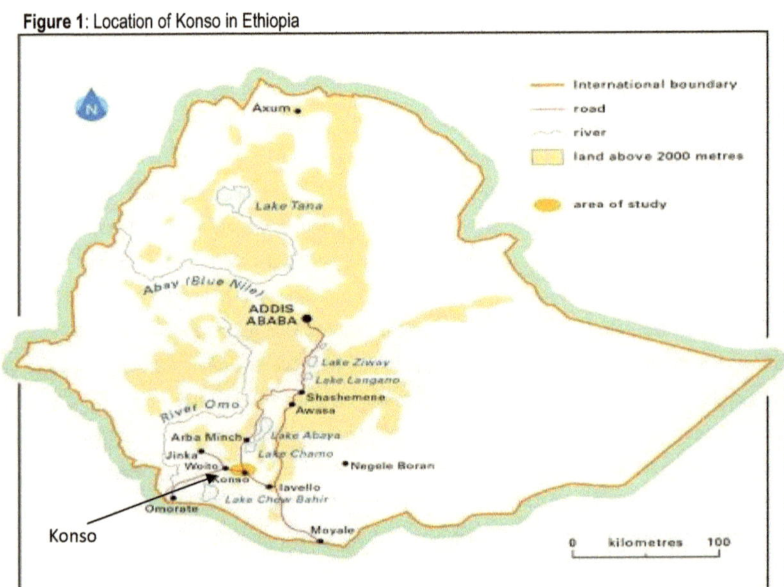

Source: http://www.geog.cam.ac.UK/research/project

Given this brief background about Konso, this essay aims to answer the following key questions: What is happing to the land in Konso woreda and why? What are the consequences of the state of the land degradation on the livelihoods of the people and ecosystem services? What is being done to solve the problem and how effective is it? In order to answer these questions, different books; journal articles; reports; electronic sources; and other relevant materials were reviewed. To reach at sound upshot, writer personal observation and experience is incorporated, as well. The remaining part of this essay includes a brief literature review, conceptual framework, DPSIR analysis of the woreda, conclusion and recommendations.

3. A Brief Literature Review

3.1. Definition of Key Concepts

3.1.1. Land

Land is one of the scarce invaluable resource without which life is impossible. According to UNEP (1992), land is a broader concept that includes climate and water resources, landform, soils, and vegetation (grassland resources and forests). FAO defines land as 'a delineable area, encompassing all attributes of the biosphere immediately above or below the earth surface, including the soil, terrain, surface hydrology, the near-surface climate, sediments and associated groundwater reserve, the biological resources, as well as the human settlements pattern and infrastructure resulting from human activity' (FAO, 1998:31). These definitions show that land is a broader concept that covers a wide range of resources (such as soil, water, and vegetation) upon which all agricultural activities and humans livelihood entirely depends. It can also be understood as a system that is made up various parts (soil, water, vegetation) working together as a whole to provide ecosystem services without which humans can't stay alive.

3.1.2. Land Degradation

Land degradation is a highly complex phenomenon and defined in various ways. For example, UNEP defines land degradation as short-term or long-term declining of the productive capacity of land (UNEP, 1992). FAO defines it as the reduction in the capacity of the land to perform ecosystem functions and services that support society and development (FAO, 2004). It is the collapse of production capability of land in terms of loss of soil fertility, soil bio-diversity and degradation of natural resources (FAO, 2002). These definitions and that of the land indicated above imply that land degradation has diverse forms including soil degradation, water degradation, vegetation degradation, and biodiversity degradation. These forms of land degradation are high interrelated. For instance, deforestation due to high demand for firewood could lead to soil erosion which in turn could result in reduction of quantity and quality of water and biodiversity. So in this essay, I will focus on these interactions in identifying and analyzing the drivers, pressures, state, impact of and response to land degradation in Konso woreda

3.2. Major Cause of Land Degradation

The causes of land degradation are multifaceted and varied. Scholars have made attempt to identify the causes of land degradation and come up with assorted factors that lead to the problem. Some of them noted that land degradation is caused by the imbalanced interaction between the natural ecosystem and human social system (Berry, 2003). Others, like World Meteorology Organization (2005) and Mulugeta (2004) classified the causes into biophysical factors (e.g. inappropriate land use) and socioeconomic factors (e.g. population growth, subsistence agriculture, poverty, illiteracy, etc).

3.3. Impact of Land Degradation

As indicated above, land degradation is complex phenomena which is manifested in many different ways: hastened soil physical, chemical and biological dwindling, detrimental changes in ecosystem services, reduction in productivity of desired plants (vegetation diminishes), dehydration of water courses, thorny weeds prevail in pastures, soils happen to be slim and rocky, etc (Berry 2003). Gradually, land degradation through these manifestations can adversely affect ecosystem services and hence human livelihood. For example, it leads to socioeconomic problems such as food insecurity, relentless poverty, poor health condition of the people, reduced livelihood opportunities. It also has adverse impact on natural environment such as decline in ecosystem resilience and provision of ecosystem services (Bossio *et al.* 2004).

4. Conceptual Framework

Understanding the nature, causes, and impacts of land degradation is highly complex due to the fact that it involves interaction and interdependence between and/or among many reinforcing factors. The DPSIR framework has thus been one of the most widely used conceptual framework to understand the complexity of the interaction of environmental issues in general and of the land degradation in particular (Svarstad et al. 2008). This framework is commonly used to demonstrate ecosystem based natural resource management (Alamz Nebyou, 2010). As can be seen from Figure 1 below, the framework has five components: driving forces, pressures, state, impacts, and responses. It shows, a series of underlying relations between driving forces (which may be due economic and human activities), pressures due to human intervention, state of land degradation in the Woreda' and its impact on ecosystems and human livelihoods, as well as responses (measures taken to deal with the problem). The framework is guided by three key questions: what is happening to the land (state) and why is this happening (driving forces and pressures)? What are the consequences for the environment and people? and what is being done and how effective is it?

In this context, driving forces are needs – desire to meet human's basic necessities and to improve people's living standard. These are the factors that cause pressure on the land resources. Pressures are human activities that are induced by driving forces and directly affect the quality of land resources (the state). Pressures can be manifested in the form of excessive, inappropriate, and change in land use. State is about the quality of land and its various components (soil, water, vegetation etc.) in terms of the functions that these components fulfill. Impact is the effect of the state of land degradation on the ecosystems functions and eventually on the livelihoods of the society. Responses are the measures taken or efforts made (mitigation and/or adaptation) by stakeholders in reaction to impacts manifested in various forms (Peter, 2004).

Source: Adapted from IEA Training Manual & LADA Project (FAO)

5. What are the Drivers of Land Degradation in Konso?

5.1. High Population Growth

In the past, the Konso people were living in valley floors but as a result of diseases such as malaria and for the purpose defending themselves from enemies, they gradually moved to mountainsides while cultivating the bottom of the hills. Eventually, as population started to grow, they commenced to think creatively about how they are going to sustainably live and produce adequate living in these mountainous areas and valleys without disturbing the environment. As a result, they came up with innovative soil and water conservation technology that is handed down from generation to generation and enabled them to live in today's marginal environment. However, for the last four decades, population of Konso has been growing very fast and becomes one of the challenges of development as well as key drivers of land degradation in the woreda.

Table 1: Trend of Population Growth, Population Density, and Land Holding

Year	No of Population	Total Area (in Square Kilometer)	Population Density (people per square kilometer)	Arable land (in hectares)	Arable land[2] (hectares per person)
1970	60,000	2354	26	153,643	2.5
1984	94,719	2354	40	153,643	1.62
1994	157,585	2354	67	153,643	0.97
2007	234, 987	2354	100	153,643	0.65
2013	280,000	2354	119	153,643	0.55

Source: Compiled from various sources (Hallpike,C. R., 1972; CSA, 1996; KDA, 2003; CSA, 2007)

Table 1 above shows that the population of Konso has been increasing in an alarming rate. The population is almost quintupled within about forty five years. Moreover, the number of people per square kilometer had increased from 26 in 1970 to 119 in 2013 showing that the number of people per square kilometer is almost quintupled as well. Population growth has also created pressure on land and resulted in shrinking of landholdings from where it was 2.5 in 1970 to 0.55 hectare per person in 2013. This implies that population growth has created profound pressure on land where the given land area can no longer sustain the live of the entire population. In general, in Konso increasing population growth has led to shrink in landholdings and resulted in over-cultivation and over use of land resources and hence decline in land productivity. These factors subsequently contributed to low crop production and heightened the problem of food insecurity (poverty). McCausland (2010) had similar observation who concluded that despite their tough character and all their hard work, due to population growth, Konso is suffering from food insecurity and environmental degradation where the average family plot shrunk down to half a hectare.

5.2. High Demand for Land Resources

The demand of the population or individuals for land resources has been increasing from time to time. Particularly need for more food, energy, income, and shelter encouraged the community to expand farmland, cut trees for firewood and charcoal making[3] and construction materials. The need to feed many mouths as a result of population growth forced the community to convert pasture land (reserved to livestock in low lands), forest land, and marginal land (kept for wild edible food and medicine) to farm lands. Beside these activities, the need to produce more food on less and fragile land for highly increasing population has reduced fallow periods. All these activities and practices

---

[2] Includes land under temporary crops, temporary meadows for mowing or for pasture, land under kitchen gardens, and land temporarily fallow. Land abandoned as a result of shifting cultivation is excluded. (http://data.worldbank.org/indicator/AG.LND.ARBL.HA.PC)

[3] Charcoal making is a new phenomena in Konso where people use it as one of the livelihood strategies to diversify their income

have led to deforestation, soil erosion, low soil fertility, loss of biodiversity and ultimately resulted in low productivity and so food insecurity and poverty.

5.3. Poverty

In Konso, poverty has encouraged poor people to overuse whichever land resource available for their exploitation. Poor people cultivate marginal lands, cut trees in the marginal areas, and ultimately degrade the land. Poverty is acting as both the driver and consequence of land degradation in the woreda. It is a driver of land degradation in that poor people in the woreda have no alternative except over using the land resources such as forests and marginal lands to survive. Particularly, poor people spent majority of their time in collecting firewood from marginal areas, cutting trees for charcoal making, and cultivating highly susceptible areas. Vegetation in highly prone areas were cut down for construction materials as well as fodder and sold in the market. Attempt of the poor to survive via all these activities and practices have resulted in reduction of the quality of the land productivity (i.e. land degradation). Low land productivity means less food and income which in turn resulted in poverty.

5.4. Current Generations' Work Ethics and Attitude

Soil and water conservation is one of the valuable cultural values of the Konso community. However, in recent years, this practice is deteriorating mainly due to current generation's work ethics and attitude. The first problem is the perception of the youth towards the practices of soil and water conservation. Some of them do not perceive terracing as one of Konso's identity. And so, their willingness to learn the skill is very low. Some of them, whom I personally met at different occasions, consider the tradition of terracing as a difficult task to act upon while few of them perceive it as backward practice. This is basically due to the fact that majority of youth are inclined to off farm and/or nonfarm activities. Even if these are among the most important alternatives that subsist to youth to break the vicious cycle of poverty, however, they have to learn the skills of maintaining this precious value. Nowadays, elders everywhere in Konso are disquieting about this situation and looking for new generation to learn from them. My concern here is that, if the current situation continuous, this valuable culture of Konso will be either ignored or abandoned and the state of land degradation could be severed.

5.5. Motivation to Expand Territories (Informal Institutions)

In Konso, each Kebele has its own cultural boundary which is also recognized by the government. At Kebele level both traditional and government administrations have been working simultaneously. Traditional administrative organs (Xella[4], Senkeleta[5], Apa Paleta[6], Apa Kanta[7], Apa timba[8], and Nama Dawra[9]) are highly influential in managing land

---

[4] The warrior group of community that serves as defense force or police force within kebele. This group is responsible to protect the boundary and people of its own kebele from enemies.
[5] Coordinator/leader of a particular warrior grade
[6] The father of the village elected from the elder's council
[7] The father of sub-village (ward)
[8] Father of the drum – a holder of scared drum

resources within boundary of respective Kebeles and in resolving conflicts caused due to the land usage. Some Kebeles fought each other over boundaries several times. The key issue here is that trees were cut down before the conflict (to trigger it) and during the conflict. The result is destruction of both human lives and plants (deforestation). In some cases, population pressure has served as a cause of conflict where Kebeles with small land size try to transcend to neighboring Kebele which relatively have large land size. The large landholding Kebele judges this as an invasion and engages in conflict. This in turn has led to destruction of trees and resulted in deforestation and land degradation.

5.6. Cultural Practices

Here, I would like to focus on Konso's land heritance system. According to the Konso culture, only the elder son of each household is a candidate to retain ownership of the original homestead and loin share of the land resource. Girls are entirely excluded from owning these resources. Regarding the Cadets, when they got married or want to marry, they are expected to move out of the original homestead. Kebele leaders (both government and traditional) are expected to give them new land for the purpose of building houses. Likewise, cadets are expected to go to find farm land and clear forest for cultivation. All of these practices have resulted in high pressure on land resource and led to land degradation via deforestation and soil erosion.

5.7. Climate Change (Rainfall Variability, Drought)

Konso community's effort in soil and water conservation is highly challenged by climate change which is resulted in rainfall variability and drought. Since 1950s konso has been one of drought prone woredas supported by many national and international humanitarian institutions. For the most part, rainfall variability has been a key problem that leads to drought and food insecurity. Figure 2 below shows annual seasonal rainfall variability (deviation from the mean) of Konso woreda from 1985 to 2010. As can be seen from the Figure, between 19982-1988, the rainfall was below the normalized average (i.e. 0 mm) and showed some improvements between 1989 and 1991 where it jumped above average (and reached at least 100 mm). From 1992 to 2000, the rainfall varied as a minimum of each year where it went up one year and went down in subsequent year. Since then, except 2010 where the rainfall reached about 150 mm (the highest in two and half decades), it fell below 100 mm and the normalized average (0 mm). These imply that the rainfall is highly varied in terms of occurrence and distribution as well as low in terms of magnitude and exposed the community to frequent drought.

---

[9] Peacemaking councilor/mediator that is responsible to resolve conflict through cultural rituals

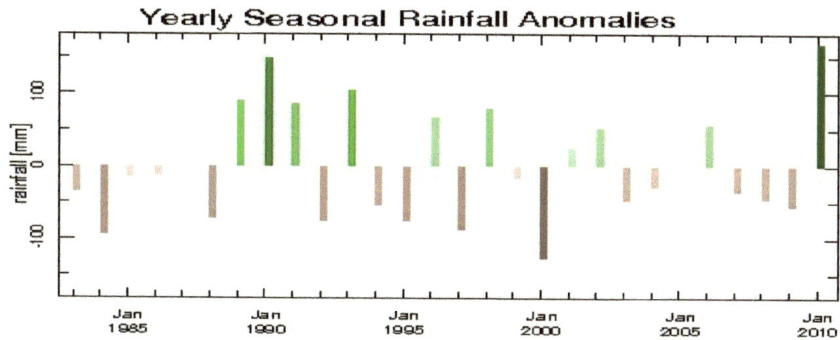

Source: Ethiopian Meteorology Agency

Even if drought has a long history in Konso, 1983/84 was the period where Konso was sternly knocked by the problem. Since then, drought has occurred at most every three years although 2000 was the most horrible year where Konso was registered as one of the worst drought hit areas in the country. The Konso people have been using different labels to describe the severity and magnitude of drought. Table 2 below shows different names given to drought in various periods.

Table 2: Occurrence of Drought in Konso

Name of the Drought	Period
Tawso	Before Emperor Menilk II
Lota Bura	Before Emperor Menilk II
Lota Bula	Before Emperor Menilk II
Herma Qalto	At the time of Emperor Menilk II
Tekela Harmasho	At time of Italian invasion
Lota Zeyse	At the time of Emperor Hailselassie
Qara Sherto	1957/1958
Lota Kewso	1974/1975
Lota Kochoro	19 83/1984
Lota Farm	1998/1999
Lota Farm	2003/2004
Lota Farm	2007/2008

Source: Adapted from Menfese, 2010

Due to unpredictability of rainfall and frequent drought, majority of Konso considered the problem as an expression of anger of God in response to their sin (Menfese, 2010). Frequent drought and erratic rainfall have exposed the land to erosion which reduced the capacity of the soil to produce even in the rainy season. During drought times, starving people are left with no alternative except destroying the land resources via search for firewood, wild animals and plants that are safe for human consumption. This condition exacerbates the problem of deforestation and result in soil erosion, decline in water resources and soil fertility. During drought episode where pasture and water is extremely scarce, livestock movement in searching for these resources is common among konso community. This problem leads to extensive grazing of a specific area or fragile land (highly steep slope areas) which in turn results in soil compaction, soil erosion, and low soil productivity.

5.8. Formal Institutions

Because of strong traditional institutions and their powerful desire to preserve boundaries between Kebeles, local government is failed to implement land use policy. As a result, Kebeles with small land size are forced to cultivate marginal lands within their boundary. This practice has led to soil erosion and degradation. The practice of paying compensation for those who are displaced from their land for development purpose (rural road, schools, and health centers construction) or for expansion of towns is either weak or non-exist in the woreda. This practice has also led many people to destroy the environment for survival purpose. The failure of government to provide alternative source of energy to rural community has also exacerbated the problem of land degradation in the woreda. Moreover, inability of local governments to create an environment for traditional leaders to actively participate and cooperate in planning and implementing integrated land management programs could be one of the drivers of land degradation.

6.   What are the Pressures that Cause Land Degradation in Konso?

6.1. Over Intensive Land Use

Beside the admirable practice of soil and water conservation; topography, population pressure, and scarcity of land have forced Konso community to extensively plough their land for several hundreds of years. This practice has led to land exhaustion and resulted in deterioration of soil fertility and its productivity. Farmers called the land 'old' when its productivity is highly declined or 'dead' when the land is not producing. To enhance soil fertility, the community has been applying various traditional strategies such as manuring and composting, intercropping, agro forestry, etc. However, as explained above, many driving forces have created pressure on the land and complicated the problem. And so, farmers are forced to look for an alternative for 'old' or 'dead' land. The primary alternative is searching for 'new or young' land wherever it is available. And so, farmers are forced to convert some protected areas to farm land which is resulted in land degradation via deforestation and soil erosion. Moreover, over intensive grazing mainly due to conversion of grazing land into farm land and increased livestock size has led to diminution of flora coverage which in turn results in soil erosion and eventually in land degradation.

6.2. Deforestation and Removal of Natural Vegetation

Konso was once blessed with fertile soil and clothed with abundant forests. However, trees were cut down to build houses, for energy purpose, and to expand farm lands. As a consequence, the soil is washed by rain and wind and left some areas of Konso with rocky skeleton (see Figure 4 below). For example, Konso Agricultural Office reported that forest land is now diminished to a total area of less than 4203ha (1.8%) (KSWARDO, 2008). Conversion of the already fragile forests to farm land, excessive collection of firewood and construction materials, the practice of charcoal making, excessive grazing in specific area, lack of appropriate technology for energy and cattle feeding, have created high pressure on the forest land and resulted in deforestation. Yilma has similar observation and concluded that the abundance of native trees and shrubs has been disappearing from forest and farmlands. This is due to the conversion of forested land into farmland and to make way for crop production (Yilma, 2006).

Figure 4: Scene of Some of the Degraded Areas

Source: Own Photo, 2010

6.3. Soil Erosion

Konso people are well known in preserving their soil and water. However, due to factors such as population growth, poverty, drought, socioeconomic activities, etc soil is under high threat. To cope up with the problems of poverty and drought, the community has cleared forest land either to generate income via sales of forest products or to expand farm land. On the other hand, shortage of land accompanied by population growth has led to over intensive land cultivation and resulted in soil erosion and low soil fertility. More than 60% of the farmers cultivate their land continuously without fallowing, the result being a steady decline in soil fertility (KDA, 2003).

6.4. Reduction of the Quantity and Quality of Water Resource

High population growth, high demand to meet basic needs and improve living standards as well as natural processes such as change in temperature, erratic rainfall, topography of the woreda, and drought have been key drivers that created pressure on the water resource. Almost all of the elders that I met in different Kebeles starting from my childhood to present in Konso said that before 50 years or so, there were a lot of springs of water around villages, every mountain had fountains of water, and rivers were perennial which had flown throughout the year. Unfortunately, this situation has been history and in recent years water is generally scarce in Konso. Almost all springs and rivers were dried up and became memory. Except Waitto, other perennial rivers have been changed in to non - perennial and the typical examples are Delibena and Segen Rivers (see Figure 5 and 6 below). Water volumes as well as discharge in seasonal rivers and streams are declined to a terrible level.

Figure 5: Delibena River during dry season   and   Rainy Season

Source: Beza Nigusie, 2007

Figure 6: Segen River during dry Season   and   Rainy Season

Source: Beza Nigusie, 2007

7. State of Quality of Land Resource in Konso

In Konso, Land is owned by both private and community and is used mainly for farming, grazing, forest, and settlement. Land use pattern of the woreda is generally categorized as follows: 38% (88,730ha) used for crop cultivation, 26% (60710ha) for grazing, 1.8% (4203ha) forest, 12% (28020ha) uncultivated and 22.2% (51837ha) others. Soil in the mid altitude area is characterized by low fertility and is very shallow less than 5cm in depth. This is mainly due to continuous cultivation over a period of hundreds of years (KSWARDO, 2008). More than 60% of the farmers cultivate their land continuously without fallowing, the result being a steady decline in soil fertility (KDA, 2003). In the low land plains, there is vast alluvial soil with high fertility and good potential for crop cultivation although moisture stress is the major constraint (KSWARDO, 2008). In Konso, almost all springs and rivers were dried up and became reminiscence. As far as forest resource is concerned, it is diminished to the point where it is hard to glimpse it. In Konso, pressure on land is very high (mainly due to driving forces mentioned above) where about 87% of households cultivate less than one hectar (KDA, 2003), see table 1, as well. Particularly, frequent drought and erratic rainfall have exposed the land to erosion which reduced the capacity of the soil to produce even in the rainy season. In general, due to the driving forces and pressures that these forces created on land, the quality of the land resources has been declining in a shocking rate. As a result, land degradation is becoming a horrendous phenomenon in the woreda.

8. What are the Impacts of Land Degradation in Konso?

8.1. Impact on Ecosystem Services

Due to low fertility of soil, scarcity of water, rainfall variability, and frequent drought, agriculture sector is failed to provide food to the community. In the past, Konso community had used edible fruits, seeds, leafs, and roots of untouched forested areas in low land and trees and shrubs around mountainsides as a source of food during drought times. But today due to conversion of forest land to farm land, the community is not able to get such service from the ecosystem. Konso people are also well known in preparing traditional medicine from vegetations. Today due to disappearance of forest, they are travelling hundreds of kilometers outside the woredas to get such services. As indicated above, water that is essential for survival is reduced in terms of quantity and quality. Moreover, due to deforestation, it is hard to find firewood without which food preparation is impossible. Likewise, due to land degradation the ecosystem is unable to provide regulation services. For example, the capacity of the ecosystem to regulate excess runoff water (hydrological service) due to deforestation, overgrazing, and soil erosion as well as regulation of water during dry seasons is highly deteriorated. Furthermore, soil covering, water and nutrient holding, soil formation, and carbon cycling capacity of ecosystem (soil services) is greatly diminished. Biodiversity services, such as pest and disease control- above and below ground, pollination, habitat services for wild animals and soil organisms is also reduced to terrible level. For example, before 20 years or so Konso is well known for honey production. Farmers frequently harvested honey both from high and low lands and sold it to merchants who came from different parts of the country. Today, due to conversion of forest land to crop land, let alone harvesting honey, it becomes hard to get a tree that could be used to dangle a beehive on it. Wild animals are completely disappeared from the woreda due to deforestation.

If business as a usually practice continues, the Konso cultural landscape which is recently registered by UNESCO will be in danger. As explained above, the practice of maintaining terraces has been weakened due to changing life style of the youth as well as population pressure. This condition could adversely affect the cultural landscape which has been attracting many national and international tourists. This could also adversely affect the tradition of skill transfer to young generation as well as national and international communities.

8.2. Impact on Livelihood

8.2.1. Impact on physical Asset

Houses, farm equipments, and livestock are among the key physical assets of Konso people. As indicated somewhere in this essay, land resources have been key path via which people enhance their physical assets. Due to low fertility of the soil, farmers could not produce more agricultural products and sale their surplus to enhance their income. Moreover, due to disappearance of forest and water resources, farmers could not get ecosystem services (e.g. honey, firewood, timber, etc) that would enhance their earnings. Due to the fact that in Konso house construction entirely depends upon natural resources (mainly vegetation), the disappearance of vegetation hampered the cadets to construct their own house and elders to maintain existing homesteads. This shows that due to lack low productivity of land resource, the community couldn't manage to generate income and buy farm equipments, livestock, and construct good houses. In general, land degradation has reduced the capacity of local community to build up their physical capital.

8.2.2. Impact on Financial Asset

Low income resulted from low agricultural productivity (crop, honey, livestock) due to land degradation via deforestation, low soil fertility, drought, etc has led to low saving. Lack of saving and physical asset in turn has resulted in low access of people to credit and led them to poverty. This situation forced majority of the community to trap in vicious circle of poverty.

8.2.3. Impact on Natural Asset

Because of drivers explained above, there has been high pressure on the land resource which is resulted in reduced farm, posture, and forest lands; low soil fertility, shortage of water resource, and loss of biodiversity. This shows that the natural asset of the community is highly deteriorating.

8.2.4. Impact on Human Capital

Human capital mainly constitutes education and health of the community. One of the clear impacts of land degradation is on the health of people. When ecosystem failed to provide provision services such as food (in the form of variety of nutrition), water (in the form of safe drinking water), shelter (that protect people from natural as well as manmade hazards), energy (that is essential for cooking and killing some germs), it is obvious that community could be vulnerable to different diseases (resulted from malnutrition, water borne diseases, malaria, and other communicable diseases). Poverty and drought has also eroded the capacity of the community to stand firm against different diseases as well as ability to pay for medical service in case of sickness. In the same token, due to low income (poverty) that is resulted from low agricultural productivity and drought, the people are either unable or unwilling to send their children to schools rather they send them to bring in some additional income that could support their livelihood. This situation has led to low enrolment ratio and high dropout rate and resulted in low literacy rate.

8.2.5. Impact on Social Capital

Social capital includes formal and informal networks, associations, and organizations of the community as well as trust and norms that guide these institutions. Social capital can facilitate or impede access of the people to various resources and services (land, water, technical assistance, market, information, financial institutions, health, education, etc). As clearly explained above, poverty is acting as driving force and consequence of land degradation in the Woreda. Poor people are usually marginalized by the community and have low access to social services such as health, education, financial, training, information, and water. For example, due to lack of collateral, a poor could not have access to financial services. Due to low income, a poor cannot pay for health, implicitly he/she cannot have access to health services. On the other hand, social capital has been playing key role in acting as a safety net scheme to the poor during drought events. It is one of the values of the Konso people that those relatively haves have a social responsibility to support those who haven't during horrifying events.

9. What Is Being Done and How Effective Is It?

9.1. Soil and Water Conservation

To deal with the problem of land degradation, the People of Konso have been practicing well organized and innovative adaptation strategies in the form of indigenous soil and water conservation. They have integrated agroforestry, terraces, and other soil and water conservation practices which have been model for the global community. These practices have been effective and enabled the community to manage and live in marginal environment. But as indicate above, the practices are now under threat which requires the attention of all stakeholders mainly of the Konso people (that is primarily responsible to maintain its identity – soil and water conservation).

9.2. Livelihood Diversification

Livelihood diversification is one the core adaptive strategies in maintaining sustainable livelihoods (Menfese, 2010) via broadening the living portfolio of the community. In Konso, besides farm activities, people are struggling to diversity their livelihood portfolio by engaging in both off-farm and non-farm activities. These activities are common during weeding and drought periods. And so, people are exchanging their labor for farm activities (individually or in group) within or outside the woreda. Some of them are generating income from environmental resources by selling firewood, charcoal, building materials, fodder, medicine, etc as well as from livestock fattening. Currently, trading and small businesses have been common and expanding to every corner of the woreda and changing the life style of the community. Seasonal migration has also been one of the livelihood strategies of Konso community where people temporary move to Karat (capital town of Konso) or neighboring woredas or zones or regions in searching for jobs. Productive Safety Net Program is supporting the livelihood of the community via cash or food for work arrangement as well. All of these activities are very important in improving the livelihood of the people and reducing their pressure on the natural resources. I feel that most of livelihood diversification strategies are effective but require strong planning and follow up system which is very weak in the Woreda.

10. Where is the Woreda Heading? (Conclusion)

The Konso people have strong mindset of 'we can' and converted this attitude to reality by effectively managing the natural environment within which they are living. However, their mindset and courage of controlling the natural environment have gradually been eroding due to low productivity of land, unpredictability of rainfall, and drought. This situation has disheartened them regarding the future of their agriculture and livelihoods. They started to externalize the cause of the problem (mainly of the rainfall variability and drought) to the supernatural being (God). This shows that people have accepted the truth that the problem is beyond their capacity and cannot be solved without the intervention of God (according to them, God is punishing them for their sins). Therefore, unless otherwise, the current trends of drivers, pressures, and impacts of land degradation on ecosystem services and livelihoods, as well as weaknesses observed in responses are either halted or improved, the woreda will be changed into bare desert where it could be impossible to live (for both current and future generation).

11. What Actions Could Be Taken for a More Sustainable Future? (Recommendations)

The practical action(s) to be taken to halt the current state of land degradation and transform the woreda towards green economy is to follow the path of Sustainable Development. This can be achieved by effectively adapting and implementing **Climate-Resilient Green Economy (CRGE) Strategy** of the country. The following issues which are relevant to the context of Konso and directly taken from the CRGE strategy could assist the woreda to transform itself to green economy via managing its land in an integrated manner.

1. **Agriculture:** Improving crop and livestock production practices for higher food security and farmer income while reducing emissions

   - Despite hesitancy about the likelihood of the agriculture sector in the woreda, as the leading occupation, agriculture will continue to be a vital sector to ensure food security for the greater part of Konso people. Integrating intensification and diversification strategies for agriculture is thus fundamental for improving land productivity of the woreda. Agriculture should be intensified through usage of improved inputs, technologies, techniques, and better residue management so as to improve land productivity and reduce the requirement for additional agricultural land that would primarily be taken from forests. Diversification which has been one of the indigenous adaptation strategies of the people should be integrated to intensification strategy and complement it in such a way that enhance land and agricultural productivity and reduce emissions

   - New agricultural land in low land areas should be created through strengthening small and medium irrigation (mainly around Woyto and Segen rivers) to reduce the pressure on forests if expansion of the cultivated area becomes necessary

- To increase the productivity and resource efficiency of the livestock sector: animals value chain efficiency as well as their health should be improved; range lands should be managed properly so as to increase its carbon content and improve the productivity of the land

2. **Forestry:** protecting and re-establishing forests for their economic and ecosystem services, including as carbon stocks. In this area the following actions should be taken:
    - Reduce demand for fuel wood through the dissemination and usage of fuel-efficient stoves and/or alternative-fuel cooking and baking techniques (such as electric, solar, or biogas stoves) leading to reduced forest degradation
    - Increase afforestation, reforestation, and forest management to increase carbon sequestration in forests and woodlands, to allow the forestry sector to store more carbon in growing forests than emitted from deforestation and forest degradation, and to reduce soil erosion. Strengthen the indigenous agro-forestry practices of the people by supporting the activities via improved inputs and technologies in both individual and communal land.
    - Trees which have market value should be expanded by both individuals and groups in order to enhance the income and reduce pressure on forest land. Particularly, micro and small enterprises should be established to process and add value on the production of *Moringa oleifera* (which is common cabbage tree called *Shelkahatta* in local language) and eventually grow to agro-processing industry that could enhance the income of community.
    - Promote area closure through rehabilitation of degraded pastureland and farmland, leading to enhanced soil fertility and thereby ensuring additional carbon sequestration

3. **Rural Electrification:** expanding the accessibility of electricity to rural areas (from hydro or solar sources) – that could be used as a source of energy for cooking, for grain mill (reduce CO_2 emission), and simply life (particularly of women and children) and improve the health of community

4. **Off-farm and non-farm activities** should be managed by a dependable government organization. Trainings that could enhance the skills of farmers, artisans, businessmen should be systematically planned and implemented. The impact of seasonal migration on family, farm activity, soil and water conservation (land management in general) should be scientifically studied.

References

Alamz Nebyou (2010). Applying the DPSIR Approach for the assessment of alternative management strategies of Simen Mountains National Park Ethiopia, A thesis submitted for the partial fulfillment of the requirements of Master of Science in Mountain Forestry, Vienna, Austria.

Bai, Z.G., Dent, D.L., Olsson, L., and Schaepman, M.E. (2008). Global assessment of land degradation and improvement: Identification by remote sensing. Report 2008/01, ISRIC – World Soil Information: Wageningen.

Berry, L. (2003). Land degradation in Ethiopia: its impact and extent in Berry L, Olson J. and Campbell D (ed): Assessing the extent, cost and impact of land degradation at the national level: findings and lessons learned from seven pilot case studies. Commissioned by global mechanism with support from the World Bank.

Bossio, D., Noble, A., Pretty, J., and Vries, F (2004). Reversing land and water degradation: Trends and'Bright Spot' Opportunities. Paper presented at the SIWI/CA Seminar. Stockholm, Sweden.

Brabant, P. (2010). A LAND DEGRADATION ASSESSMENT AND MAPPING METHOD: A standard guideline proposal. Les dossiers thématiques du CSFD Issue 8, French Scientific Committee on Desertification. Les Petites Affiches (Montpellier, France).

Central Statistical Authority (1996). Population and Housing Census, 1994. Addis Ababa, Ethiopia.

Central Statistical Authority (2008). Statistical Abstract, January 2007. Addis Ababa, Ethiopia

FAO (1998) Terminology for Integrated Resources Planning and Management. retrieved from https://www.mpl.ird.fr/crea/taller-colombia/FAO/AGLL/pdfdocs/landglos.pdf on January 10, 2016, Rome

FAO (2002) Land Degradation Assessment in Dry lands (LADA) Project: Meeting Report, 23-25 (World Soil Resources Reports)

FAO (2004). Methodological Framework for Land Degradation Assessment in Dry lands (LADA). Land and Water Development Division, Rome

(FAO) (2011). The state of the world's land and water resources for food and agriculture (SOLAW): Managing systems at risk. FAO, Rome and Earthscan, London.

Girma Taddese (2001). Land Degradation: A Challenge to Ethiopia. *Environmental Management*, 27(6), 815–824.

Hallpike, C. R. (1972) *The Konso of Ethiopia: A study of the values of a Cushitic People.* Oxford University Press

Jolejole-Foreman, M. C., Baylis, K., and Lipper, L. (2012). Land Degradation's Implications on Agricultural Value of Production in Ethiopia: A look inside the bowl, Selected Paper prepared for presentation at the International Association of Agricultural Economists (IAAE) Triennial Conference, Foz do Iguacu, Brazil, 18-24 August 2012

Konso Development Association-KDA (2003). Socio-economic Profile of Konso Special Woreda, Konso Karat

Konso Special Woreda Agriculture and Rural Development Office (2008). General Physical Features of Konso Special Woreda, Konso Karat

McCausland A. (2010). Permaculture in Konso & Ethiopia. Konso Karat

Menfese Tadesse (2010). Living with Adversity and Vulnerability: Adaptive Strategies and the Role of Trees in Konso, Southern Ethiopia, Doctoral Thesis Submitted to Swedish University of Agricultural Sciences, Department of Urban and Rural Development, Uppsala

Mulugeta, L (2004). Effects of land use change on soil quality and native flora degradation and restoration in the highlands of Ethiopia. Implication for sustainable land management. Ph.D Thesis. Swedish university of Agricultural Science. Uppsala, Sweden.

Nachtergaele, F., Petri, M., Biancalani, R., Van Lynden, G., and Van Velthuizen, H. (2010). Global Land Degradation Information System (GLADIS). Beta Version. An Information Database for Land Degradation Assessment at Global Level. Land Degradation Assessment in Drylands Technical Report, no. 17. FAO, Rome, Italy.

Peter, K. (2004). The DPSIR Framework: Paper presented on workshop on a comprehensive /detailed assessment of the vulnerability of water resources to environmental change in Africa using river basin approach. UNEP Headquarters, Nairobi, Kenya

Svarstad, H., Petersen, L.K., Rothman, D., Siepel, H., and Wätzold, F. (2008). Discursive biases of the environmental research framework DPSIR. *Land Use Policy* 25, 116-125

Temesgen Gashaw, Amare Bantider and Hagos G/Silassie (2014). Land Degradation in Ethiopia: Causes, Impacts and Rehabilitation Techniques. *Environment and Earth Science*, 4(9)

UNCCD (2013). A Stronger UNCCD for a Land-Degradation Neutral World, Issue Brief, Bonn, Germany

Thiombiano, L., and Tourino-Soto, I. (2007). Status and Trends in Land Degradation in Africa. In Sivakumar V. K. M. and Ndiang'ui Nd. (Eds.), *Climate and Land Degradation*, 39-56), Retrieved from http://www.researchgate.net/publication/226165171 on January 1, 2016.

UNEP (1992). World atlas of desertification. Edward Arnold. London.

World Meteorological Organization (2005). Climate and land degradation retrieved from https://www.wmo.int/pages/themes/wmoprod/documents/WMO989E.pdf on January 5, 2016

Yilma Sunta (2006). The Role and Status of Women in the Food System of the Konso of Southwest Ethiopia, a PhD Thesis in Social Anthropology, Department of Sociology and Anthropology, Addis Ababa University,

YOUR KNOWLEDGE HAS VALUE

- We will publish your bachelor's and master's thesis, essays and papers

- Your own eBook and book - sold worldwide in all relevant shops

- Earn money with each sale

Upload your text at www.GRIN.com and publish for free